Vania Castro, V.P.P
José Firmino Nogueira Neto

Lipid profile and US-CRP in an interlaboratory approach

Vania Castro, V.P.P
José Firmino Nogueira Neto

Lipid profile and US-CRP in an interlaboratory approach

Inter-laboratory comparison between methodologies and sample stability in lipid profile dosage and US-CRP

ScienciaScripts

Cover image: www.ingimage.com

This book is a translation from the original published under ISBN 978-613-9-70484-2.

Publisher:
Sciencia Scripts
is a trademark of
Dodo Books Indian Ocean Ltd. and OmniScriptum S.R.L publishing group

120 High Road, East Finchley, London, N2 9ED, United Kingdom
Str. Armeneasca 28/1, office 1, Chisinau MD-2012, Republic of Moldova, Europe
Printed at: see last page
ISBN: 978-620-8-21105-9

DEDICATORY

I dedicate this work to the most important people in my life, for what they have taught me and passed on to me, for their unconditional and unceasing support, for who I am. To my parents, my husband, my daughter, my siblings, my family and my friends.

ACKNOWLEDGEMENTS

To the Sacred Heart of Jesus for accepting all the suffering that has come from this battle and for giving me the strength and courage I need for this victory.

To my husband Jorge João dos Santos Castro Filho for his understanding in times of absence and constant encouragement, as well as to my angelic daughter Maria Beatriz.

To my parents for having been at the forefront of my educational process, always guiding me along good paths.

To my siblings, friends and other relatives who believed and through dialogue, thoughts and prayer helped me to complete this journey.

To my advisor Prof Dr José Firmino Nogueira Neto for supporting me when I decided to do my master's degree and to stay in this postgraduate course amid so many difficulties.

To the members of the Fragility in Elderly Brazilians - FIBRA project, Prof Dr Roberto Alves Lourenço, for allowing the use of the data and for understanding that investing in the quality of services and processes is fundamental.

To my colleagues at the Lipid Laboratory - LabLip for all their affection in helping me with this work.

To the professors of the Professional Master's Programme in Laboratory Medicine and Forensic Technology (MPSMLTF), and to the coordinator, Professor Luis Cristóvão de Moraes Sobrino Pôrto, who shared his knowledge.

The scientist is not the man who provides the real answers; he is the one who asks the real questions.

Claude Lévi-Strauss

SUMMARY

Introduction: According to O'Kane et al, in 2008 it was shown that 88.9% of laboratory errors occur in the pre-analytical phase. In the clinical analysis laboratory, the quality control system can be defined as all the systematic action necessary to provide confidence and security in all tests and prevent errors from occurring. Storage time can vary from days to months or even years, influencing the definition of storage temperature. Long-term storage can result in inadequate cryopreservation for certain analytes and can denature lipoproteins. Objectives: The aim of this study was to evaluate the pre-analytical phase, internal quality control and the effect of freezing (-80°C) on the storage time of blood serum and plasma collected with ethylenediamine tetraacetic acid (EDTA). Material and method: The dosages were carried out at the Lipid Laboratory - LabLip, and stored at - 80 °C for three years. They were then returned to the Clinical Pathology Service of the Piquet Carneiro Polyclinic at UERJ using the same methodologies and the results were compared. Results: The lipid profile (TC, HDLc and TG) and US-CRP of 103 samples were analysed, 73 in serum and 30 in plasma in two highly qualified laboratories. Discussion: After re-dosing, lower results were found for HDLc and TC respectively (correlation coefficient in serum 0.48 and 0.62) paired t-test in serum (TC p 0.0012 and HDLc p 0.0001). Conclusions: The data obtained in the evaluations of the results from different laboratories and storage times revealed that when the samples were stored for a long time after redosing in serum, they showed differences in certain analytes, such as TC and HDLc, in which significantly decreased results were obtained, unlike in plasma, which after three years of storage at -80°C was redosed and applied to the paired t-test, the analytes TC and PCR-US maintained stability.

Keywords: Comparability. Methodology. Storage. Stability.

SUMMARY

INTRODUCTION

Clinical analysis laboratories provide health care and play an important role in clinical decisions. With scientific technology, their complexity has also increased and laboratory processes have been modified, unifying the benefits of information technology and being impacted by varying levels of automation. Thus, laboratory testing also takes place in a complex environment, where procedures, equipment, technology and human knowledge coexist, with the aim of providing laboratory findings for care, diagnosis, prognosis, therapeutic monitoring and scientific production.[1]

O'Kane et al, in 2008, showed that 88.9% of laboratory errors are made in the pre-analytical phase.[2] Every laboratory analysis aims to obtain results that are compatible and reliable with the methodology used; however, various factors can cause different values to be obtained for a given laboratory analysis of the same biological material.[3]

The quality control programme is a system that manages and regulates the conditions necessary for quality improvement to be implemented and maintained, thus achieving the client's quality and granting the laboratory reliable results, resolving non-conformities and promoting improvements in the system.[3-4]

The aim of this study was to gather and organise knowledge about the pre-analytical phase, quality control, comparison of methodologies between two clinical laboratories and expected results per group of individuals in the Fragility in Elderly Brazilians - FIBRA II project.

FIBRA Project

The aim of the study was to determine the risk profile and factors associated with frailty in community-dwelling elderly people. The source population consisted of individuals aged 65 and over, living in the northern neighbourhoods of the city of Rio de Janeiro, Brazil, and clients of a health insurance company. The study was a cross-sectional, cohort baseline study, with a sample stratified by gender and age, with n=213 participants undergoing clinical analyses. The probability of repeated hospitalisations (PIR) screening tool was used for risk stratification. Logistic regression analysis was carried out to study the association between PIR and a set of sociodemographic, health status, functional and cognitive variables, after bivariate analysis. 6.7 per cent of elderly people were found to be at high risk of hospitalisation. The risk of hospitalisation was associated with cancer, falls, chronic obstructive pulmonary disease and medication used, as well as the following conditions: receiving a visit from a health professional, having been bedridden at home, living alone and practising activities of daily living.[5]

Target audience

A cross-sectional, descriptive study was carried out on the baseline population of the "Fragility in Elderly Brazilians - FIBRA" study. The study population and sample were made up of individuals aged 65 or over, living in neighbourhoods in the north of the city of Rio de Janeiro, who were part of the client register of a health and pension foundation for federal civil servants and their dependents, with coverage in various Brazilian municipalities. The geographical delimitation was defined by the FIBRA-RJ project coordinators for logistical convenience.[5]

The sample was selected using the stratification process, based on the crossing of the age and sex variables, forming ten natural strata, made up of individuals aged 65 and over up to 100 years old, divided into ten-year bands. Each final stratum was obtained through reverse random sampling, maintaining the proportions of the source population strata, except for individuals aged 95 and over, all of whom were interviewed. The sample size was calculated in such a way that the coefficient of variation in each natural stratum was 15% for proportion estimates of around 0.07 with a 95% confidence level. [5]

Elderly people who needed a substitute informant because they had any of the following conditions were excluded from the study: cognitive impairment - defined by a score of less than 12 on the *Mini-Mental State Examination* (MMSE); sensory deficits that compromised communication and reading; terminal illnesses of any kind. Among other limitations, the substitute respondent would not answer the self-assessment of health, an item that makes up the instrument used to stratify the risk of frailty. [5]

Recruitment took place by telephone between 5th January 2009 and 13th January 2010.[5]

The study was approved by the Research Ethics Committee of the Pedro Ernesto University Hospital of the State University of Rio de Janeiro (1850-CEP/HUPE). All participants signed an informed consent form.[5]

Data collection and the research instrument were carried out at home through a single interview, lasting approximately 90 minutes, with a questionnaire of structured questions and measures of physical, functional and mental performance. [5]

The questionnaire was made up of sociodemographic data, such as marital status and housing, schooling, skin colour/race, personal income, age, gender and availability of a carer in case of need, and data on health status, through self-rated health and self-reported chronic diseases, such as systemic arterial hypertension (SAH), coronary heart disease, diabetes mellitus, cancer, arthropathies, chronic obstructive pulmonary disease, osteoporosis, stroke, hearing or visual impairment, falls in the last year, and urinary and faecal incontinence. The individual answered a questionnaire on the use of health services in the last year, characterised by the number of hospitalisations and length of stay, number of medical consultations, need for home visits by health professionals, need to remain bedridden due to illness and number of medicines used regularly in the three months prior to the interview, smoking and whether they practice physical activity.[5]

Anthropometric measurements were taken, such as weight and height, and performance tests, such as gait speed - the average of three assessments of the time taken to walk 4.6 metres in a straight line - and handgrip strength - the average of three measurements taken by a Jamar dynamometer (SAEHAN Corporation, Yangdeok-

Dong, South Korea) on the dominant upper limb.[5]

The probability of repeated hospitalisation for each participant was calculated using the probability of repeated admission (PRA), made up of eight items included in the questionnaire: (1) self-rated health - with the following response options: "very good, good, fair, poor or very poor"; (2) hospitalisation in the last year; (3) number of medical consultations in the last year; (4) diabetes mellitus; (5) coronary heart disease; (6) gender; (7) availability of a carer in case of need; (8) age.[6] The logistic equation and the regression coefficients corresponding to each of the eight items were described by Pacala et al[7] In the present study, the name of the instrument, in Portuguese, was changed to aPIR.[8]

Pre-analytical phase

The pre-analytical phase comprises all the steps that take place before the test is carried out, including the request, preparation of the individual, collection, identification of the samples, and their - . 9
handling and processing.

In the pre-analytical phase of laboratory care, a lot of data is highly relevant, such as gender, age, use of medication, health insurance and others. The use of computerised systems, which are widespread in laboratory environments, allows an additional series of information to be collected, in relation to the time and date of the service, post 9
collection, operator, among others.

The use of automated equipment with bidirectional interfacing (in which the equipment is fed with the results of the analyses, receives work instructions from the laboratory system and returns results to this same system) can allow access to another set of information produced in the analytical process. In addition, automating the process of obtaining results, acquiring and storing data makes the volume of information more reliable, agile and easy to handle. [10]

The doctor requesting the test and his assistants don't always instruct the individual on the test and the collection of the sample, so the laboratory must provide guidelines for each type of test, and the phlebotomist must be concerned about compliance with the technical requirements of the collection and the potential biological risks. Likewise, the people responsible for packaging, preserving and transporting the sample must ensure the safety and integrity of the material and themselves. [11]

When blood is taken for laboratory tests, it is important to control and avoid certain variables that can interfere with the accuracy of the results. Pre-analytical conditions include chronobiological variation, gender, age, position, physical activity, fasting, diet and the use of drugs for therapeutic or non-therapeutic purposes. There are other conditions that must be taken into account, such as the contemporaneous performance of therapeutic or diagnostic procedures, surgery, blood transfusion and infusion of solutions. [12]

Laboratory process

The process involves requesting the test, orientation, preparation of the individual, collection, processing of the samples, the analysis itself, expression of the results and their interpretation in comparison with reference values and the clinical situation of the individuals studied. [13,14]

Work instructions (Procedures):

Blood collection is an important factor in the diagnosis and treatment of various diseases. [15]

The collection materials used to obtain the blood are critical, as they have an impact on the test results.

Work instruction (FIBRA II project sample collection sequence):

a) Tube containing sodium citrate;
b) Tube with clot activator, with gel to obtain serum;
c) Tube containing EDTA.

Work instructions (Order of collection procedures):

a) Reception of the individual;
b) Analysing the request;
c) Checking information: name, date of birth, gender, clinical indication, use of medication, etc;
d) Check that the individual is fasting properly for the requested test and if they are allergic to latex, if so, use a tourniquet and latex-free gloves.

Work Instruction (Venipuncture Procedures):

a) Separate the necessary and appropriate material for collecting the requested tests;
b) Label the tubes and record the time of collection. Some laboratories also identify the phlebotomist by putting on gloves;
c) Positioning the individual;
d) Apply the tourniquet, select the puncture site and the vein to be punctured;
e) Antisepticise the puncture site and wait for the antiseptic to volatilise;
f) Perform the venepuncture and when the blood flow starts, ask the individual to open their hand;
g) Loosen and remove the tourniquet;
h) Complete the collection tubes in the correct order;

i) Place a gauze pad over the puncture site;
j) Remove the needle and activate the safety device;
k) Apply pressure to the puncture site until the bleeding stops, then apply the dressing;
l) Check that there is special processing for the tests collected;
m) Send the labelled tubes to the laboratory.

Possible errors:

a) Incorrect identification, sample exchange, haemolysis, homogenisation, centrifugation at the wrong rpm, inadequate storage, errors in the use of anticoagulants;
b) Samples with inadequate identification should not be processed;

Collection instructions:

a) Preparation of the individual;
b) Material to be collected;
c) Collection time;
d) Effective identification of the individual;
e) Correct identification of the sample collected;
f) Special care;
g) Recording the identity of the sample collector or receiver;
h) Safe disposal of the material used in the collection;
i) Correct completion of the individual's registration;
j) All samples must be labelled so that they can be traced if necessary;
k) The samples must be stored for the specified time and at the specified temperature under conditions that guarantee the stability of the properties for carrying out and repeating the analyses;

l) Consecutive freezing and defrosting is not permitted;
m) The quality of the biological specimen is of paramount importance to the success of the analysis;
n) Make sure that the containers are tightly closed and that the contents are not leaking;
o) Place the tubes or vials containing the biological material in a plastic bag or jar in an upright position before placing them in the cooler;
p) Important: The thermal case must contain a quantity of dry or recyclable ice compatible with the quantity of material being sent and a thermometer for temperature control.

Pre-analytical variables

Pre-analytical variables have a major impact on the quality of laboratory results and are classified into three categories: physiological variables, specimen collection variables and other interfering factors, which can lead to erroneous interpretations of test results. [16]

When the results are analysed in the laboratory, there are some changes linked to physiological variables such as gender, age, race, pregnancy, etc. The extent of the changes in these substances depends on the diet and the time elapsed between ingestion and sample collection. Foods rich in fat increase the concentration of triglycerides in the body. Diets rich in proteins and nucleotides, on the other hand, promote increased levels of ammonia, urea and uric acid. The effect of physical exercise, smoking and alcohol use, interference linked to altitude, among others, on test results is also well known. [16]

There are other pre-analytical factors such as: collection variables that have tourniquet time as an agent, blood collected from venous access sites with drug infusion, among others. [16]

There are other interfering factors, such as prolonged contact time of the serum or plasma with the cells, the existence of haemolysis to varying degrees, haemoconcentrations caused by evaporation, incorrect sample storage temperature, incorrect transport, incorrect use of additives (anticoagulants). [16]

The serum or plasma should be separated from the blood cells as quickly as possible. If you exceed the maximum time of two hours, some analytes may be interfered with.

Temperature is also important for the viability of the sample. The ambient temperature in the laboratory is considered to be between 22 and 25°C.[18]

Refrigerating the sample at temperatures between 2 and 8°C inhibits cell metabolism and stabilises certain thermolabile constituents.[19]

Samples must be transported in isothermal, hygienisable and waterproof cases, when required. They must be labelled with biological risk symbols. [19]

Quality control

Laboratory analyses tend to obtain results that are compatible with the methodology used. However, various factors can lead to different values for a given laboratory analysis of the same biological material.[20] Quality in the laboratory has evolved considerably in recent years and quality control is part of a broader programme. Quality includes the processes of management, improvement and quality assurance, 2221
and quality control is included in continuous quality improvement.

The components of the quality control programme are: sample quality; standard operating procedures (SOPs); technical quality assurance of employees; quality control maintenance and records; analysis results; participation in external quality monitoring programmes; laboratory safety standards; reliability of equipment performance; and quality assurance of laboratory materials.

Description of LabLip's Quality System

LabLip participates in two Quality Control programmes at international and national level. The External Quality Control Programme (PCQE) is made up of control samples that are supplied monthly in a control *kit* to LabLip for analysis as a Proficiency Test. The purpose of these analyses is to assess the analytical processes carried out under the same conditions as the analyses of the samples from the various projects we carry out. Through the results obtained from participating in these programmes, LabLip ascertains precision and accuracy.

LabLip's PCEQ comprises analytical evaluation through the precision of the analytical system, using internal control samples (intra-laboratory) Internal Quality Control (IQC) and evaluation of analytical accuracy with control samples analysed (between laboratories) External Quality Control (EQC).

The CQE samples are analysed and the results found by LabLip are evaluated by the programme provider in the form of a periodic report that allows us to take action for continuous improvement. At the end of a cycle of participation, we have achieved proficiency through a Certificate of Excellence. Since the beginning of our participation in these programmes, we have obtained an EXCELLENT rating from both.

The CQI is carried out with control samples supplied by the National Quality Control Programme (PNCQ) and also controls supplied by a highly qualified company with all the requirements. Both supply lyophilised human serum for the internal control of all biochemistry analytes using the same batch of control samples, allowing the system to be checked over a period of time without changing the biological control material. The tests are carried out, analysed and approved before each dosage that makes up the projects LabLip is involved in.

Controls, calibrators and their functions

Control samples with known values are analysed daily to assess the accuracy of the tests. With this efficient and reliable way of carrying out laboratory procedures, we obtain valid results that contribute to clinical diagnosis and the various lines of research in which we participate. The purpose of CQI is to guarantee reproducibility (precision), check the calibration of analytical systems and indicate when corrective action should be taken when non-compliance arises.

Calibrator Serum is freeze-dried bovine serum added to various components until it reaches levels suitable for calibrating automatic analysers. It contains no preservatives that could interfere with biochemical determinations.

To interpret the results of the internal control samples, LabLip uses two control samples at different concentration levels: human control I (normal) and human control II (pathological), so that the information is valid for checking that the desirable control levels are maintained, with the aim of monitoring analytical performance. The Levey-Jennings Control System, expressed as a graph, is also used as a quality assurance tool. Each analyte has a graph, and the acceptable limits are respected, given by the standard deviation in relation to the mean obtained after a minimum of dosages according to internationally accepted protocols referenced in bibliographies.

Equipment

The analytical system comprises the measuring equipment and instruments, which are the biochemical analysers and those supporting the tests (centrifuges, pipettes, etc.). All the equipment and instruments used by LabLip are checked every three months for preventive maintenance by specialised technicians and representatives of the manufacturer, and corrective maintenance is carried out when necessary. If necessary, any components that show signs of jeopardising the results are replaced. Performance is rigorously observed through post-maintenance tests for insertion back into the technical execution of the analyses.

Technical staff

The Analytical Quality Control Programme implemented at LabLip also includes the technical professional who carries out the analyses. To this end, there are systematised hierarchical levels for decision-making, in accordance with the competences, qualifications and qualifications described in internal quality documents. Periodically, and whenever necessary, LabLip promotes training through a continuing education programme and checks its effectiveness.

Final considerations

Quality management is very important, given the credibility crisis associated with the area. It is not common for the public service to hire a quality programme for accreditation, due to the resources, because it only works if everyone is aware of it. Total Quality Management (TQM) thus emerges as an instrument around which institutions can be restructured to meet the real health needs of the country.[22]

QM, based on participatory management, has expanded management work, promoting decentralisation and making control more efficient. This gave management a new role, that of change agent, advisor and educator, which helped to maintain a more participatory relationship. [22]

Adding value based on awareness of systems knowledge, scenario studies and continuous learning has become the differentiating element. [23]

With the implementation of QM, gains were made in terms of human resources, guaranteeing internal customer satisfaction in their working environment. The customer's needs have been met, society has been recognised and there have also been changes in the hospital's statistical indicators. [23]

It has been confirmed that QM has achieved satisfactory results. The credibility of the management model is therefore obtained, demonstrating that its applicability guarantees the satisfaction of customer needs as well as those of the organisation's members. [23]

Lipids and US-CRP

Cardiovascular diseases are the main cause of death, especially atherosclerosis. This disease is considered to be an active and recurrent inflammatory state that affects peripheral and central blood vessels and its development is related to the presence of risk factors.

Risk factors for atherogenic development include dyslipidaemias, hypertension, diabetes, smoking, physical inactivity, family history and metabolic syndrome. Diagnosis, monitoring and treatment of dyslipidaemias are mainly based on blood concentrations of desirable and altered values for these serum lipids, based on the III Brazilian Guidelines on dyslipidaemias. [24]

Lipid profile values, total cholesterol (TC), triglycerides (TG) and high-density lipoprotein cholesterol (HDLc) are references for assessing cardiac risk, diagnosing dyslipidaemias and monitoring treatment. Therefore, the choice of methods used to measure blood lipids is extremely important. [24]

The Atherosclerosis Department of the Brazilian Society of Cardiology, in view of the wide range of scientific publications on the treatment of dyslipidaemias and the prevention of atherosclerosis and the importance of their impact on cardiovascular risk, brought together a committee of experts to present the updated Brazilian Guidelines on Dyslipidaemias and the Prevention of Atherosclerosis, published in the Brazilian Archives of Cardiology in October 2013.[24] High cholesterol is the main modifiable risk factor based on scientific studies. It is therefore coherent that reductions in cholesterol, especially LDLc levels, through lifestyle changes and drugs, over time, have great benefit in reducing cardiovascular outcomes. [24]

CRP belongs to the pentraxin family of proteins[25] and its determination in blood samples is used as an important factor in assessing predisposing factors for cardiovascular diseases and diagnosing inflammatory states in patients with osteoarticular diseases [26]. Currently in a population aged 40-79, the distribution of serum CRP was similar in men and women after taking into account smoking and the use of hormone replacement therapy. [27] Amer et al.[28] were 222229 found increased CRP levels in healthy elderly Egyptians and Delongui et al. described that serum CRP levels ranged from <0.175-48.7 mg/L and were influenced by gender, age and body mass index (BMI) in a healthy population in southern Brazil. New methodologies and strategies for detecting and quantifying CRP are currently available which imply differences in specificity and sensitivity.[30] In addition, controls for possible confounding variables are often neglected. [30]

Therefore, the aim of this study was to assess the variations between lipid profile values and US-CRP and whether these alterations can generate different results.

CHAPTER 1

OBJECTIVES

1.1 General

Evaluation of the pre-analytical phase and internal quality control. A comparative study of biochemical tests, including lipid profile (TC, HDLc and TG) and PCR - US, carried out in two clinical analysis laboratories.

1.2 Specific

To analyse the factors associated with the pre-analytical alterations found in laboratory tests; to compare a sample of 103 results obtained in a university clinical laboratory and to evaluate the effect of storage at -80°C of the PCR-US, CT, TG and HDLc analytes.

CHAPTER 2

MATERIAL AND METHODS

This study comprised a systematic review of laboratory test changes in terms of sample storage time for lipid profiles and US-CRP and a comparison of results between two laboratories.

Research was carried out in books, technical and scientific publications and databases, including PubMed and Scientific Electronic Library Online (SciELO).

2.1. Study design

We used data from the FIBRA project carried out in a clinical research laboratory at UERJ's Faculty of Medical Sciences, LabLip, and redosage in the Cápsula clinical laboratory, which provides care to the public, both located in the Piquet Carneiro Polyclinic.

The venous blood samples were collected during the years 2010 and 2011, in the morning (from 7am to 10am), after overnight fasting, and the serum and plasma were processed on the same day. Vacuum tubes were used for collection, following the recommendations and care required to obtain samples suitable for laboratory procedures.

2.2. Inclusion criteria

A total of 103 samples were chosen for redosing, as they had sufficient volume for the analyses required. The patients evaluated had the following diseases: arthrosis, osteoporosis and cancer, and some were smokers.

2.3. **Exclusion criteria**

All samples with insufficient volume, haemolysis, lipaemia, inadequate identification and collection that did not meet the requirements laid down for this procedure in the standard operating procedure (SOP) were excluded.

2.4. **Quality controls**

Materials such as standards, calibrators, controls, reagents and supplies used in the laboratory routine were analysed, checking suppliers, method of preparation, shelf life, conservation and storage.

2.5. **Comparability**

This study compares the results of the FIBRA project between two laboratories. This study evaluated 103 samples related to the project, and was instructed on the procedures for preparing the individual for sample collection, fasting, diet prior to collection, sample handling, transport[18] , storage, disposal, among other pertinent indications. We also analysed the recommended volume of samples and the conditions under which they could become unacceptable according to the laboratory's SOP. In the pre-analytical phase, a questionnaire was administered to each individual.

Serum and plasma samples were processed to determine the analytes and then stored in aliquots in the -80°C *freezer* for three years.

Storage

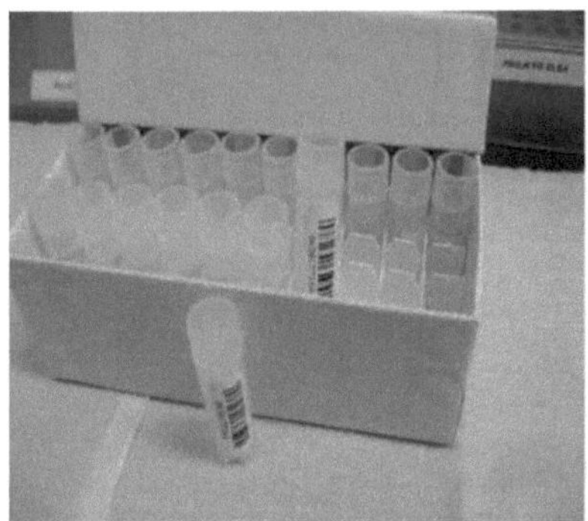

Figure 1 - Cryotubes

Source: Spinelli, 2012.

They were packed in cryotubes and stored in a *freezer* at -80°C. The facilities for storing samples locally at -80°C were designed to safely store the aliquots.

2.6 **FIBRA project**

A study was carried out to determine the risk profile and factors associated with frailty in elderly people in a community in the northern part of the city of Rio de Janeiro, Brazil. The target population consisted of individuals aged 65 and over, living in various neighbourhoods in this locality, and individuals from a healthcare provider. The study was a cross-sectional cohort study with a sample stratified by sex and age. [5]

One hundred and three samples were selected for redosing due to the sufficient volume, including 30 plasma samples and 73 serum samples.

The RIP screening tool was used for risk stratification. [5]

Logistic regression analysis was carried out to study the association between RIP and a set of sociodemographic, health status, functional and cognitive variables, after bivariate analysis. 6.7 per cent of elderly people were found to be at high risk of hospitalisation. The risk of hospitalisation was associated with cancer, falls, chronic obstructive pulmonary disease and medications used, as well as the following conditions: being visited by a health professional, having been bedridden at home, living alone and practising activities of daily living. [5]

2.7 **Analysis procedures**

LabLip Laboratory: The biochemical tests for total cholesterol and triglycerides were carried out using the oxidase and peroxidase method, HDLc using the direct detergent method and PCR- US using the turbidimetry method and high sensitivity latex. These analytes were measured on an automated A25 photometric reading device, brand Biosystems, in accordance with the manufacturer's instructions and laboratory protocols. The performance of the analytical process was assessed by a quality control system using materials provided by the PNCQ, Rio de Janeiro, Brazil and Prevecal, Spain.

Figure 2 - Automated random access A25 analyser used for biochemical dosages using spectrophotometry and turbidimetry methods

Cápsula Laboratory: The biochemical tests total cholesterol and triglycerides were carried out using the enzymatic colourimetric method, homogeneous enzymatic colourimetric HDLc, PCR high sensitivity turbidimetry method. The analytes were measured on a Cobas Integra 400plus automated device, which was used to carry out the following methodologies: photometry, turbidimetry, polarised fluorescence and ion-selective electrode potentiometry.

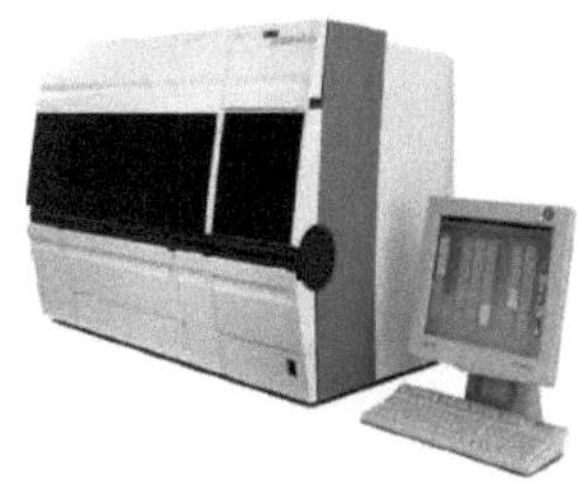

Figura 3 - Cobas Integra 400plus, automated random and continuous access analyser with integration of four measurement principles

ControlLab's external quality control programme was used to evaluate the analytical results.

Although the commercial names are different for each manufacturer, the *kits* refer to the same methodological principle.

Table 1 - Description of LabLip's quality system

Analect	**CQI**	**Manufacturer**	**EQC**	**Methodology**
PCR-US	Protein control serun I			Turbidimetry
CT	Human Contour I (normal) and Human Contour II (pathological)	Biosystems	Prevecal PNCQ	Spectrophotometry
HDLc	Lipid Control serun I			Spectrophotometry
TG	Cont. human I (normal) and			Spectrophotometry
	human cont. II (pathological)			

Source: The author, 2015.

Table 2 - Description of Cápsula's quality system

Analect	**CQI**	**Manufacturer**	**EQC**	**Methodology**

PCR-US	Precinorm protein	Roche	ControlLab	Turbidimetry
CT	Precinorm U plus and precipath U plus			Enzymatic colourimetric
HDLc	HDLc Precinorm			Enzymatic colourimetric
TG	Precinorm U plus and precipath U plus			Colourimetric enzymatic

Source: The author, 2015.

2.8 **Statistical analysis**

The EPI info and Excel programmes were used for statistical analysis of the data.

CHAPTER 3

RESULTS

The table below shows the results that were dosed at LabLip (2010 and 2011) and re-dosed at Cápsula (2014).

Table 3 - Representing serum dosage results

NIC	TC (mg/dL) LabLip	TC (mg/dL) Capsule	TG (mg/dL) LabLip	TG (mg/dL) Capsule	HDLc (mg/dL) LabLip	HDLc (mg/dL) Capsule	PCR-US (mg/dL) LabLip	PCR-US (mg/1) Capsule	OBS
6963	232	196	129	93	47	32	0,29	2,1	
7039	213	264	171	211	39	32	0,24	1,8	
7077	137	121	102	83	54	44	0,15	8,2	
7078	193	224	126	120	76	66	0,32	2,6	
7079	221	270	116	128	47	42	0,42	4,3	
7080	192	182	64	62	59	46	0,22	3,7	
7081	143	171	74	124	65	37	0,26	0,5	
7134	217	208	129	131	64	41	0,34	2,2	
7136	245	237	191	176	49	37	0,24	1,0	
7137	241	316	204	269	53	39	0,17	1,2	
7139	151	135	103	95	58	44	0,2	0,9	
7182	233	226	144	113	78	53	0,17	1,1	
7184	244	215	108	92	45	23	0,1	0,4	
7188	202	203	100	124	73	61	0,25	2,2	
7189	217	203	131	136	51	34	0,08	1,8	
7190	170	171	128	172	54	37	0,05	2,3	
7197	131	117	187	125	35	25	0,05	3,6	
7202	179	210	46	109	88	39	0,07	4,9	
7203	199	196	55	92	105	42	0,09	0,8	
7204	188	207	66	126	70	66	0,08	1,3	
7212	246	202	347	217	42	24	0,17	1,0	
7213	234	255	275	259	59	36	0,24	1,7	
7214	169	173	113	108	45	32	1,73	12,4	
7251	203	209	61	71	65	43	0,19	1,0	
7252	224	248	148	163	38	28	0,11	0,4	
7259	220	220	86	111	67	44	0,27	1,3	
7302	178	203	160	192	50	35	0,9	7,3	
7303	158	134	79	80	58	41	1,84	16,0	
7386	224	244	182	166	60	42	0,26	2,2	
7390	183	217	199	193	50	38	0,43	4,0	
7399	120	124	99	116	39	35	0,33	3,0	

7446	210	217	78	84	46	31	0,81	7,7	
7449	268	219	486	358	36	13	0,21	1,8	
7492	237	219	209	166	41	23	0,3	1,2	H+
7500	135	112	156	122	47	31	0,15	0,6	

Legend: haemolysate (H).

Source: The author, 2014.

Table 3 - Representing serum dosage results

7501	287	248	87	161	67	33	0,06	1,6	
7503	168	138	120	106	41	28	0,21	1,5	
7515	188	210	129	132	60	38	0,46	6,5	
7523	136	141	88	88	32	25	0,27	2,0	
7524	212	224	129	123	54	35	0,07	0,5	
7525	194	243	123	136	56	46	0,33	2,6	
7526	178	194	169	158	35	26	0,12	0,9	
7527	142	234	133	179	44	51	0,44	5,3	
7528	169	189	186	185	44	29	1,56	14,2	
7578	159	211	89	104	59	55	0,11	0,9	
7585	151	193	59	81	45	42	0,12	1,1	
7586	204	212	56	69	57	52	0,14	1,3	
7597	249	256	208	189	34	18	0,22	1,8	
7599	190	153	128	99	57	32	0,08	0,5	
7600	178	186	66	68	53	43	0,1	0,7	
7602	224	243	166	155	51	27	0,4	3,4	
7605	190	200	94	94	61	48	0,09	0,5	
7607	231	280	102	118	72	62	0,18	1,4	
7608	173	181	182	174	51	38	0,38	2,4	
7609	257	267	110	115	66	52	0,49	3,8	
7619	182	185	66	67	63	53	0,4	3,2	
7620	226	246	297	297	59	40	0,13	0,9	
7625	181	204	89	86	59	50	0,20	1,3	
7804	256	281	111	121	60	44	0,11	0,7	
7916	226	247	167	161	54	37	0,50	2,7	
7917	206	226	57	59	57	53	0,51	4,3	
7918	223	215	272	269	42	27	0,13	1,1	
7977	169	192	168	169	49	33	0,35	2,9	
8060	180	218	99	105	64	57	0,12	0,9	
8066	165	180	123	122	43	33	0,18	1,4	
8067	259	289	137	142	67	50	0,89	8,6	
8068	228	251	112	119	47	31	1,29	16,1	
8069	183	206	92	98	77	67	0,26	2,1	
8070	145	158	122	118	62	56	1,27	12,5	
8071	175	188	140	145	53	37	1,29	13,6	
8072	168	180	129	124	52	36	0,85	7,5	
8077	211	224	222	218	51	35	0,46	4,2	
8078	172	185	189	189	42	29	0,15	1,1	

Source: The author, 2014.

Table 4 - Representing plasma dosage results

NIC	TC (mg/dL) LabLip	TC (mg/dL) Capsule	TG (mg/dL) LabLip	TG (mg/dL) Capsule	HDLc (mg/dL) LabLip	HDLc (mg/dL) Capsule	PCR-US (mg/dL) LabLip	PCR-US (mg/1) Capsule	OBS.
7076	190	157	139	108	52	32	0,15	0.5	
7082	282	246	135	101	53	36	0,27	1.9	
7132	235	233	87	88	90	71	0,98	5.8	
7181	276	247	215	159	47	29	1,78	15.3	
7183	180	144	54	43	74	57	0,70	5.6	
7187	188	146	94	94	65	52	0,2	0.9	
7195	156	152	146	125	41	28	0,92	7.2	
7196	224	221	161	139	48	35	0,01	3.1	
7249	207	207	127	103	70	46	0,29	2.2	
7250	195	211	118	119	54	47	0,1	0.2	
7253	229	235	79	91	95	70	0,09	0.2	
7258	249	219	51	73	83	53	0,15	0.7	
7260	241	208	90	95	55	32	0,58	1.6	
7266	127	139	49	63	56	36	0,11	0.4	
7387	204	215	138	130	61	40	0,57	7.6	
7395	179	219	558	554	38	15	0,16	0.7	LT
7397	172	176	152	144	43	39	0,3	2.3	
7432	173	189	209	198	69	54	0,16	0.9	
7433	193	199	85	86	53	38	0,25	2.8	
7434	293	304	236	206	48	26	0,77	9.5	
7435	214	227	68	75	87	72	0,08	0.7	
7448	233	212	139	138	39	29	0,33	2.8	
7450	239	238	206	218	44	27	0,17	1.0	
7506	234	236	203	176	56	34	0,33	2.8	
7517	100	121	109	105	45	33	0,11	0.8	
7519	178	204	173	162	48	30	0,82	6.9	
7529	147	161	104	99	43	30	0,45	3.5	
7581	210	200	177	147	34	22	1,21	9.4	
7601	154	152	126	109	33	19	1,49	16.7	
7624	160	171	174	162	55	42	0,75	6.5	

Legend: slightly turbid (LT).

Source: The author, 2014.

The results of the statistical analyses of the data are shown in graphs 1-4 with units in mg/dL, redosage in the Cápsula laboratory in 2014.

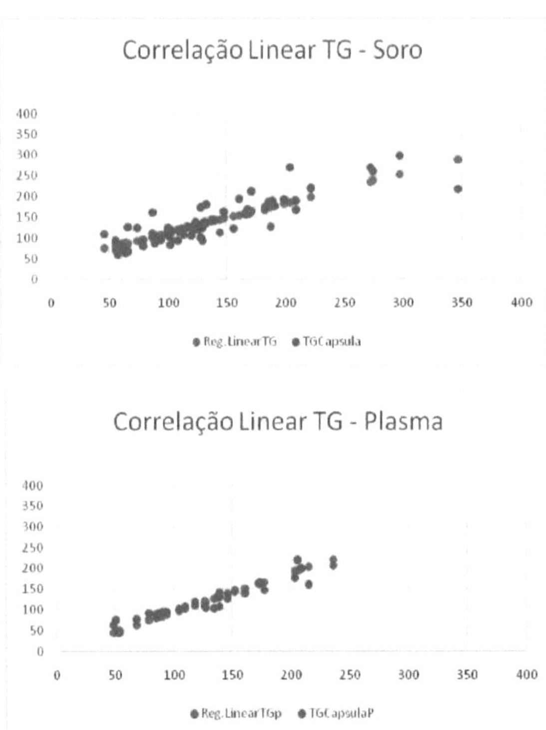

Graph 1 - Representative of the comparison of triglyceride results at LabLip and Cápsula laboratories

Source: The author, 2015.

Linear regression

TG	Serum	Plasma
N	73	30
Correlation r	0,90	0,98
Median	124	114
Correlation coefficient r^2	0,80	0,97

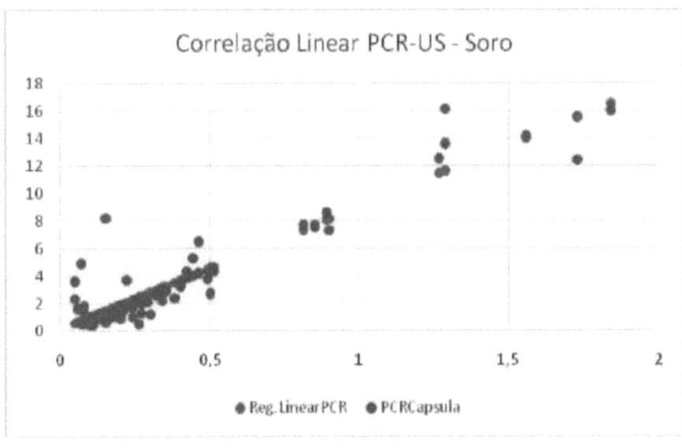

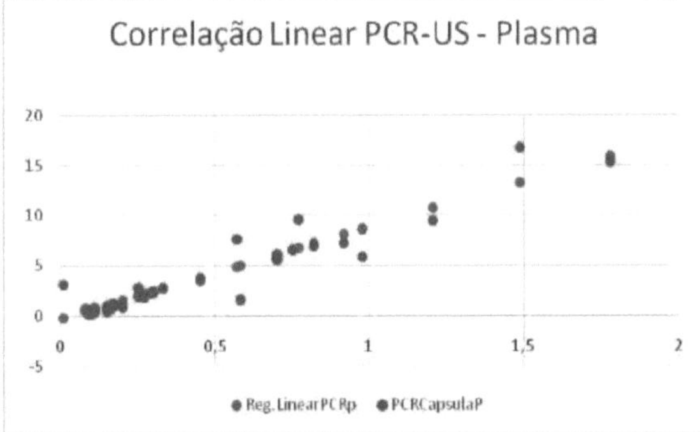

Graph 2 - Comparison of US-PCR results at LabLip and Cápsula laboratories

Source: The author, 2015.

Linear regression

PCR-US	Serum	Plasma
N	73	30
Correlation r	0,93	0,94
Median	1,8	2,55
Correlation coefficient r^2	0,86	0,88

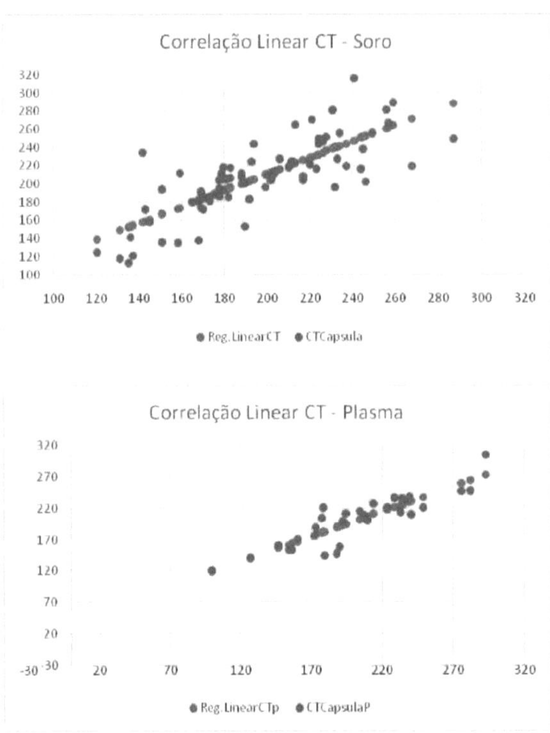

Graph 3 - Comparison of total cholesterol results at LabLip and Cápsula laboratories

Source: The author, 2015.

Linear regression

CT	Serum	Plasma
N	73	30
Correlation r	0,79	0,88
Median	209	207,5
Correlation coefficient r^2	0,62	0,78

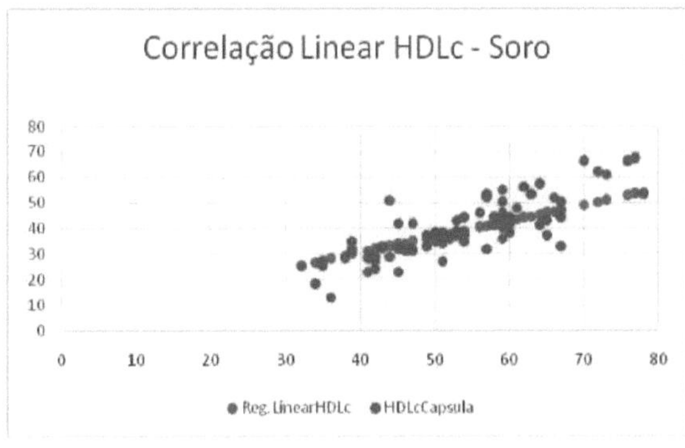

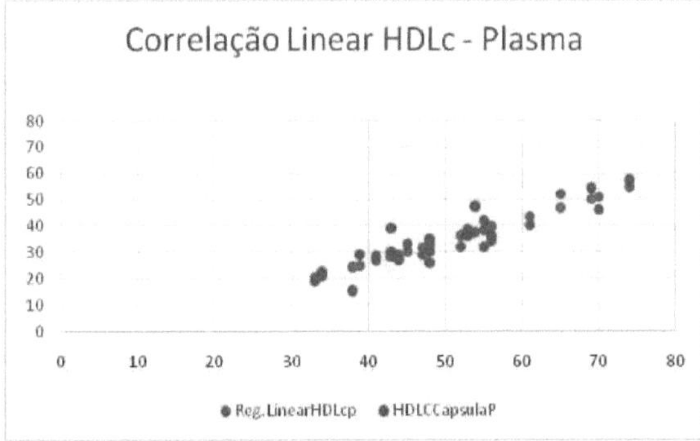

Graph 4 - Comparison of HDLc results at LabLip and Cápsula laboratories

Source: The author, 2015.

Linear regression

HDLc	Serum	Plasma
N	73	30
Correlation r	0,70	0,94
Median	38	35,5
Correlation coefficient r^2	0,48	0,89

Table 5 - *Paired t test results*

Analyses	P	T	DF	Epd
Serum				
*CT	0,0012	3,3687	72	3,066
TG	0,8252	0,2217	72	3,832
*HDLc	0,0001	13,4361	72	1,131
PCR-US	0,1207	1,5705	72	0,017
Plasma				
CT	0,5318	0,6328	29	3,845

*TG	0,0041	3,1199	29	3,120
*HDLc	0,0001	16,4397	29	1,024
PCR-US	0,0126	2,6605	29	0,028

*Statistically significant results.

Legend: two-tailed p-value (p); distribution (t); degrees of freedom (df); standard error of the difference (Epd). Source: The author, 2015.

In the US-CRP and TG measurements, the serum samples stored under freezing at - 80°C did not change over three years of storage, as their values remained stable. We obtained a correlation coefficient of: PCR-US 0.86 and TG 0.80, respectively.

We observed that there was no statistically significant difference, showing that the storage method is in adequate condition.

Plasma with EDTA is a highly flexible material for preserving these dosages, as long as the collection conditions are respected and the appropriate temperature for preservation is maintained.

As for the results found for the HDLc dosage, it was found that after freezing in the Cápsula laboratory, they showed a lower correlation coefficient: serum 0.48 than those presented in LabLip, showing significant statistical differences, confirming the information cited in the Roche *kit* leaflet, as HDLc cannot be analysed in serum in samples frozen for more than 30 days at -80°C.

The paired test showed statistically significant results in the plasma samples for the TG and HDLc analytes and in the serum for TC and HDLc, see table 5.

CHAPTER 4

DISCUSSION

According to Thorense et al 31, plasma samples stored under freezing at - 80°C did not change over 240 days of storage, as the cholesterol and triglyceride values remained stable. Studies related to our procedures show that there was no change when working with samples from Wistar rats.[32,33]

Unlike this study, which used serum and plasma in humans, CT levels also decreased after the samples were stored frozen at -80°C. After three years, they were redone and the correlation coefficient was obtained: serum 0.62, plasma 0.78.

We have verified that the information given in the Roche kit's leaflet is very reliable, as HDLc cannot be analysed in serum in samples frozen for more than 30 days at -80°C, the correlation coefficient in serum was 0.48, but in plasma it remained stable with a correlation coefficient of 0.89.

Studies in different laboratories show that the main changes that resulted in errors described in the research were storage time, with 78.6 per cent.[34]

The processes carried out in the pre-analytical phase were correct in both laboratories, and any errors in this phase can be ruled out.

The determination of serum lipids can be influenced by various pre-analytical factors. Factors related to collection (posture, tourniquet time) and obtaining, handling and preserving the sample must be carefully controlled by laboratories. There are also pre-analytical factors involved exclusively in the individual, such as physical exercise, diet, alcohol consumption, smoking, pregnancy and others. These aspects will reflect in different numerical values in the dosages.[35]

In the clinical laboratory, the quality control system component can be defined as all the systematic action necessary to give confidence to laboratory procedures in order to meet the health needs of the individual and prevent errors from occurring.[36,4,37]

Quality management is highly relevant, as it associates the difficulties in implementing quality systems in public organisations.

The QM is therefore an instrument around which institutions can be restructured to meet the country's real health needs. [22]

To guarantee quality in the release of results by laboratories, intra-laboratory and inter-laboratory controls are carried out. The intra-laboratory controls are commercial, purchased from a company recognised by ANVISA. All results are only released after checking the controls on the Levy-Jennings chart, which only allows two standard deviations up and down.

Therefore, both laboratories had adequate internal controls and equipment for the methodologies used to obtain reliable results.

The methodology used includes standardised commercial *kits* (Biosystems and Roche) for measuring total cholesterol, HDLc, PCR-US and triglycerides. The leaflets for the *kits* mention the storage time, while Biosystems does not provide any information on storage at -80°C. The analytes studied could be measured in serum and plasma with EDTA.

The principle of the TC method is enzymatic colourimetric, cholesterol esters are cleaved by the action of cholesterol esterase and produce free cholesterol and fatty acid. HDLc the concentration of HDLc cholesterol is determined enzymatically, by cholesterol esterase and cholesterol oxidase coupled with polyethylene glycol to the amino groups (around 40 per cent). [38] In TG, triglycerides are rapidly and completely hydrolysed to glycerol, followed by oxidation to dihydroxyacetone phosphate and hydrogen peroxide. The hydrogen peroxide produced then reacts with 4-aminophenazone and 4-chlorophenol under the catalytic action of peroxidase to form a red dye (Trinder end-point reaction). The colour intensity of the red dye formed is directly proportional to the concentration of triglycerides and can be determined photometrically.[39]

PCR-US is based on the turbidimetry method with particle reaction intensification.[40]

There are procedures that make storage safer, through the biobank, whose purpose is to store the collection of various types of biological materials, related to individual medical information, also involving a large number of participants involved in a study or a particular institution. [41]

The importance of the biobank is to make it possible to draft future case-cohort studies that will aid scientific procedures. [41]

CHAPTER 5

CONCLUSIONS

It was confirmed that in the set of tests, US-CRP, TC, TG and HDLc, carried out between two different laboratories and using the same methodologies, there were significant alterations in the serum samples measured for HDLc and TC. Although the triglyceride results did not change with storage time, the lipid profile was compromised, as we found significant changes in all the analyses carried out on samples stored for three years. We found that none of the studies that used frozen samples exceeded two years.[42,43,44,45,46]

The data obtained from evaluating the results from different laboratories and storage times revealed that serum samples, when stored for a long time after redosing, show differences in certain analytes, such as TC and HDLc.

It was observed that there was no statistical difference in serum in terms of TG and PCR-US in samples stored for three years, unlike the CT and HDLc analytes which, when redried in serum, showed relatively lower results. With regard to plasma, redosing all the analytes (CT, HDLc, TG and PCR-US) after three years of storage at - 80°C generated results that were applied to the *t-student* and maintained stability in the samples, with the exception of TG and HDLc, which also showed decreased results.

CHAPTER 6

REFERENCES

1- Plebani, M. Exploring the iceberg of errors in laboratory medicine. Clin Chim Acta. 2009 Mar 18. 404: 16-23.

2- O'Kane, M. , Lynch, P.L.M. , Mc Gowan, N. Development of a system for the reporting, classification and grading of quality failures in the clinical biochemistry laboratory. Annals of Clinical Biochemistry. 2008. 45(2): 129-134.

3- Stein E.A. Lipids, lipoproteins, and apolipoproteins. In: Tietz NW, ed. Fundamentals of Clinical Chemistry. 3rd ed. Philadelphia: WB Saunders; 1987:448-481.

4- Berlitz, F.A. Quality control in the clinical laboratory: aligning process improvement, reliability and patient safety. J Bras Patol Med Lab. 2010. 46(5)353-63.

5- Lourenço, R. A. Rede FIBRA-RJ: frailty and risk of hospitalisation in elderly people in the city of Rio de Janeiro, Brazil. 2014. [accessed 2014 Jan 7]. Available at: http://www.scielosp.org/pdf/csp/v29n7/12.pdf pdf.

6- Fried, L.P. et al. Frailty in older adults: evidence for a phenotype. J Gerontol A Biol SciMed Sci 2001. 56:M146-56.

7- Pacala, J.T., Boult C., Boult L. Predictive validity of a questionnaire that identifies older persons at risk for hospital admission. J Am Geriatr Soc 1995;43:374-77.

8- Katz, S. et al. Studies of illness in the aged. The index of ADL: a standardised measure of biological and psychosocial function. JAMA 1963;185:914-9.

9- Lippi G. et al. Preanalytical variability:the dark side of the moon in laboratory testing.
Clin Chem Lab Med 44 (4) 358-365, 2006.

10- Mauricio, Pacheco de Andrade. A proposed data structure for application in investigation of analytical processes in clinical laboratories. [dissertation]. Faculty of Pharmaceutical Sciences; 2007.

11- Guder, W.G, et al. Samples: From the patient to the laboratory. The impact of preanalytical variables on the quality of laboratory results. 2.ed. Darmstadt: Cit Verlag GMBH; 2001.

12- McPherson, R.A.; Pincus, M.R. Henry's clinical diagnosis and management by laboratory methods. 21.ed. Philadelphia: Saunders Elservier, 2007.

13- Valenstein, P. N.; Sirota, R. L. Identification errors in pathology and laboratory medicine. Clin. Lab. Med. 2004. 24 (4): 979-96.

14- Sirota, R. L. Error and error reduction in pathology. Arch. Pathol. Lab. Med. 2005. 129(10): 1228-1233.

15- CLSi H3-a6, Procedures for the collection of diagnostic Blood Specimens by Venipuncture; approved Standard, 6th ed.

16- Frazer, C.G. Biological variation: from principles to practice. Washington: AACC Press; 2001.

17- Brazilian Society of Clinical Pathology/Laboratory Medicine. Programme for the Accreditation of Clinical Laboratories - PALC. PALC Standard - Version 2013. [accessed 2014 Jan 16]. Available at: www.sbpc.org.br.

18- NCCLS. H21-A4. Collection, transport and processing of blood specimens for testing plasma-based coagulation assays. 2003.

19- Brazilian Society of Clinical Pathology/Laboratory Medicine. Recommendations of the Brazilian Society of Clinical Pathology/Laboratory Medicine for venous blood collection. 2.ed. Barueri: Minha Editora; 2010.

20- Moura RA, Wada CS, Purchio A, Almeida TV. Laboratory techniques. São Paulo: Atheneu; 1998.

21- Chaves CD. Quality control in the clinical analysis laboratory. J Bras Patol Med Lab 2010;46(5):352.

22- Roesh, S.M.A.; Antunes, E.D.D.; Total quality management: top-down leadership versus participative management. Rev Adm. 1995; 30(3):38-49.

23- Barbosa, A.P. Qualidade em serviços de saúde: análise dos instrumentos utilizados na promoção e garantia da qualidade na prestação de serviços hospitalares em um hospital geral de grande porte no município de São Paulo [thesis]. São Paulo: Getúlio Vargas Foundation Business School; 1995.

24- - Xavier, H.T.I., M.C. et al. V Diretriz Brasileira de Dislipidemias e Prevenção da Aterosclerose. Arq. Bras. Cardiol. 2013;101(4 Supl1): 22.

25- Maekawa, Y., T. Nagai, and A. Anzai. Pentraxins: CRP and PTX3 and cardiovascular disease. Inflamm Allergy Drug Targets. 2011;10(4): 229-35.

26- Salazar, J., et al, C-reactive protein: clinical and epidemiological perspectives. Cardiol Res Pract, 2014.

27-Ahmadi-Abhari, S., et al, Distribution and determinants of C-reactive protein in the older adult population: European Prospective Investigation into Cancer-Norfolk study. Eur J Clin Invest. 2013; 43(9): 899-911.

28-Amer, M.S., et al, High-sensitivity C-reactive protein levels among healthy Egyptian elderly. J Am Geriatr Soc. 2013; 61(3): 458-9.

29- Delongui, F., et al, Serum levels of high sensitive C reactive protein in healthy adults from Southern Brazil. J Clin Lab Anal. 2013; 27(3): 207-10.

30- Braga, F., M. Panteghini. Biologic variability of C-reactive protein: is the available information reliable? Clin Chim Acta, 2012. 413(15-16): 1179-83.

31- Thorense, S.I. et al. Effects of storage time chemistry results from canine whole blood, heparinised whole blood, serum and heparinised plasm.Vet Clin Pathol. 1998; 21(3):88-94,

32- Spinelli O. M. et al. - Effect of temperature and time on the storage of metabolites in the PLASMA of recently weaned wistar rats, Revista da Sociedade Brasileira de Ciência em Animais de Laboratório, São Paulo, Brazil.

33- Oliveira, F.S. et al. Effect of freezing and storage time of lamb blood serum on the determination of biochemical parameters. Semina: Ciências Agrárias, 2011;. 32(2):717-722.

34- Costa, V.G. et al. Main biological parameters evaluated in errors in the pre-analytical phase of clinical laboratories: a systematic review. J Bras Patol Med Lab. 2012; 48(3):163- 168.

35- IV Brazilian Guidelines on Dyslipidemias and Atherosclerosis Prevention Guideline of the Atherosclerosis Department of the Brazilian Society of Cardiology. Arq. Bras Cardiol 2007; 88.

36- Lopes, H.J.J. Quality assurance and control in the clinical laboratory. Technical and scientific advice from Gold Analisa Diagnóstico Ltda [access in 2014 May8]. Available at http://www.goldanalisa.com.br/publicacoes/Garantia_e_Controle_da_Qualidade_no_Laborato rio_Clinico.pdf

37- Motta, V.T. Biochímica clínica para o laboratório: princípios e interpretações. Caxias do Sul: EDUCS; 2003.

38- Sugiuchi H, Uji Y, Okabe H, Irie T et al. Direct Measurement of High-Density Lipoprotein Cholesterol in Serum with Polyethylene Glycol-Modified Enzymes and Sulfated a-Cyclodextrin. Clin Chem 1995;41:717-723.

39- Wahlefeld AW, Bergmeyer HU, eds. Methods of Enzymatic Analysis. 2nd English ed. New York, NY: Academic Press Inc, 1974:1831.

40- Breuer J. Report on the Symposium Drug Effects in Clinical Chemistry Methods. Eur J Clin Chem Clin Biochem 1996;34:385-386.

41- Hallmans G.,Vaught J.B. Best practices for establishing a biobank. Methods Mol.Biol. 2011; 675: 241-60.

42- Kale, V.P. et al. Effect of repeated freezing and thawing on 18 clinical chemistry analytes in rat serum. J Am Assoc Lab Anim Sci. 2012. Jul; 51(4):475-8.

43- Cuhadar, S. et al. - Stability studies of common biochemical analytes in serum separator tubes with or without gel barrier subjected to various storage conditions. Biochem Med (Zagreb) 2012; 22(2):202-14.

44- Brinc, D. et al. Long-term stability of biochemical markers in paediatric serum specimens stored at -80 °C: a CALIPER Substudy. Clin Biochem. 2012;45(10-11):816-26.

45- Tanner, M. et al. Stability of common biochemical analytes in serum gel tubes subjected to various storage temperatures and times pre-centrifugation. Annals of Clinical Biochemistry. 2008; 45(Pt 4):375-379.

46- Cray, C. et al. Effects of storage temperature and time on clinical biochemical parameters from rat serum. J Am Assoc Lab Anim Sci. 2009; 48(2):202-4.

ANNEX A - Questionnaire carried out in the pre-analytical phase of the FIBRA project

Race

1- What colour or race are you?
( )White
( ) Black
( ) Mulatto / cabocla / parda
( ) Indigenous
( ) Yellow / oriental
( ) NS
( ) NA
( ) NR

Lifestyle habits: Smoking

2- Do you currently smoke?
( ) Yes
( ) No
( ) NS
( ) NA
( ) NR

2.1- For those who answered YES, ask: "How long have you been a smoker?

2.2- For those who answered NO, ask:
( ) Never smoked?
( ) Have you ever smoked and quit?
( ) NS
( ) NA
( ) NR

Use of medication

3- How many medicines have you used regularly in the last three months, either prescribed by your doctor or on your own?
3.1- Can you use the medication correctly without help?
3.2- Are you able to use the medication, but need some kind of help?
3.3- Are you able to take your medication without help?

Exercises

4- Do you exercise? Which exercises?

Perceived physical health

5- Heart disease such as angina, myocardial infarction or heart attack?
6- Hypertension?

7- Stroke?Cerebral ischaemia?

8- Arthritis, arthrosis or rheumatism?
9- Depression?
10-Osteoporosis?
11-Cancer?

ANNEX B - FIBRA II data

NIC	Gender	Age	HAS	AVE	Dislip.	Cancer	Loc canc	Dementia	Osteo 2	From the ost	Loc ost	Pes.	Height
7506	F	98	999	999	999	999	999	1	Pododacilios	999	999	36	148
7266	M	79	1	1	1	1	999	2	999	1	999	69	162
7395	M	84	1	2	1	2	999	2	999	2	999	79	166
7527	F	85	999	999	999	999	999	1	999	999	999	999	999
7449	M	67	1	2	2	2	999	2	999	2	999	79	164
7804	F	89	2	2	1	2	999	1	999	2	999	60	158
7181	M	80	2	1	2	2	999	1	999	2	999	63	165
7182	M	69	2	2	2	2	999	2	999	2	999	70	162
7213	F	81	1	1	1	2	999	2	999	2	999	65	150
7250	F	88	1	2	2	2	999	1	999	2	999	45	148
7258	M	84	2	2	1	2	999	2	999	2	999	78	165
7259	M	81	1	2	2	2	999	2	999	2	999	75	162
7917	F	81	1	2	1	2	999	2	999	2	999	64	147
7916	M	86	2	2	0	2	999	2	999	2	999	70	166
8072	F	89	999	999	999	999	999	1	Knee	999	999	999	999
8066	M	69	2	1	0	2	999	2	999	2	999	84	170
8078	F	90	999	999	999	999	999	1	999	999	999	55	146
7625	F	87	1	1	1	2	999	1	999	2	999	999	999
7203	F	73	1	2	2	2	999	2	999	2	999	67	159
7585	M	82	2	2	2	2	999	2	999	2	999	66	175
7081	F	84	999	999	999	999	999	1	Column	999	999	46	143
7251	M	67	2	2	2	2	999	2	999	2	999	74	163
7523	M	75	2	2	1	2	999	2	Drop	1	Drop	64	165
7578	F	84	1	2	2	2	999	1	Don't know	2	999	60	165
24													
7080	F	74	999	999	999	999	999	1	999	999	999	71	160
7190	F	88	999	999	999	999	999	1	999	999	999	56	142
7136	M	96	2	2	2	999	999	1	3° left outodicle	2	999	42	148
7977	F	75	1	2	2	2	999	1	999	2	999	89	167
7529	F	97	999	999	999	999	999	1	999	999	999	999	999
7602	F	78	999	999	999	999	999	1	999	999	999	999	999
7619	F	82	999	999	999	999	999	1	999	999	999	40	147
7082	M	76	999	999	999	999	999	1	999	999	999	58	166
8068	M	91	999	999	999	999	999	1	Osteoporosis	999	999	48	165
7528	F	93	2	2	2	2	999	1	999	2	999	73	144

7620	F	71	2	2	2	1	Left breast (1997)	2	999	2	999	59	148
7446	F	84	1	2	2	1		1		1	Column,	65	154
7599	F	80	1	2	2	1	Endometrium in 2003	1	999	1		83	150
8071	F	84	1	2	2	1	Mama	2		1		57	152
7188	F	83	1	2	2	1		2	Knee, vertebral fracture, cuff	1		64	150
7387	M	70	1	2	1	1		1	999	1	Knee, lumbar, dorsal,	76	162
7302	F	78	1	2	1	1		2		1		66	152
14													
7212	M	86	1	1	2	2	999	1	Lumbar	1	Lumbar	63	173
7607	M	82	2	2	2	2	999	2	Scoliosis	1	Scoliosis	69	179
2													
7039	F	92	999	999	999	999	999	1	Arthrosis, osteoporosis	999	999	60	149
7303	F	71	1	2	1	2	999	2	Osteoporosis	1	Traumatic vertebral fracture	66	152
7253	F	85	1	2	1	2	999	2	Osteoporosis	1	Osteoporosis	43	147
7435	F	73	1	2	1	2	999	2	Osteoporosis	1	Osteoporosis	57	154
7183	F	73	2	2	2	2	999	2	Osteoporosis	1	Osteoporosis	49	156
7600	F	91	2	2	2	2	999	1	Osteoporosis of the femur	2	999	52	147
7432	F	78	2	2	1	2	999	2	Osteoporosis	1	Osteoporosis	60	142
7519	F	75	1	2	1	2	999	2	Osteoporosis	1	Osteoporosis	60	155
7132	F	80	1	2	1	2	999	2	Osteoporosis	1	Osteoporosis	40	145
7252	F	86	1	2	1	2	999	1	Knee arthritis	1	Osteoporosis	52	146
7624	F	82	1	2	2	2	999	1	Osteoporosis	0	Leg pain	72	148
7139	F	78	1	1	1	2	999	1	Osteoanthosis of the knees	1	Osteoporosis	83	162

									/ osteoporosis				
7450	F	90	1	2	1	2	999	2	Column	1	Column	70	149
8069	F	75	1	2	2	2	999	2	Osteopema	1	Osteopema	60	154
7601	M	91	2	1	1	2	999	2	Gonarthrosis	1	Legs	59	169
7500	F	73	1	1	1	2	999	2	Spine / femur	1	Lumbar / cervical	72	157
7202	F	93	2	2	1	2	999	1	999	1	Knees	999	999
7196	M	82	2	2	0	2	999	2	Right shoulder	1	Right shoulder	56	174
7079	F	78	2	2	2	2	999	1	Armo se	1	Knees	58	163
8060	F	70	1	2	1	2	999	2	Osteopenia	1	Osteopenia	65	150
7390	F	81	1	2	1	2	999	2	999	1	Knees	63	148
7501	F	68	1	2	1	2	999	999	999	1	Spine/femur	59	151
8067	F	78	1	2	1	2	999	2	Knee, lower back	1	Knee, spine	112	162
7597	F	80	1	1	1	2	999	2	999	1	Tenosynovitis in hand.	64	145
7197	F	70	1	2	1	2	999	2	999	1	Osteopenia	66	160
7189	M	80	1	2	2	2	999	2	Hand	1	Hand and foot	65	170
7077	F	73	2	2	2	2	999	1	Knee	1	Knee	61	155
7214	F	88	1	2	2	2	999	2	Knees	1	Knees	60	162
7526	F	73	1	2	1	2	999	2	Hands, shoulders, knees, spine	1	Hands, shoulders, knees, spine	65	158
7204	F												
30													
7586	F	74	1	2	2	2	999	2	OA Knees	1	OA Knees	61	149
7.515	M	97	2	2	2	2	999	1	Knees	1	Knees	43	154
7525	F	68	1	2	2	2	999	2	Arthrosis / Osteopema	1	Arthrosis of the spine/ Osteopema	72	155
7249	F	89	999	999	999	999	999	1	Arthrosis	999	999	54	141
7137	F	89	1	2	2	2	999	2	Arthrosis	1	Hand arthritis	66	156
7076	F	71	1	2	1	2	999	1	Knee arthritis	1	Knee arthritis	56	137
7581	F	91	1	1	0	2	999	1	Knee arthritis	1	Knee arthritis	999	999
7434	F	75	1	2	2	2	999	2	Arthrosis	1	Knee arthritis	93	160
7517	M	91	1	2	2	2	999	1	Arthrosis	1	Hands	64	155

									of the hands. lower back				
7187	F	82	1	2	2	2	999	2	Knee arthritis	1	Knees	46	146
8070	F	76	1	2	1	2	999	1	Knee arthritis	1	Knee arthritis	68	153
7609	F	69	2	2	0	2	999	2	Arthrosis of the spine	1	Arthrosis of the spine/kne e	56	147
7448	M	93	999	999	999	999	999	1	Knee arthritis	999	999	93	140
7184	F	90	999	999	999	999	999	1	Knee arthritis	999	999	54	143
7433	F	76	2	2	2	2	999	2A	arthritis/te ndinitis	1	Arthritis / tendonitis	61	155
7195	F	84	1	2	1	2	999	1/	rthrosis of the knee d.	1	'label. knees d.	99 9	999
7260	F	87	1	2	2	2	999	1	Knee arthritis	1	Knee arthritis	66	152
7492	F	76	1	2	1	2	999	1	Arthrosis of the spine	1	Arthrosis of the spine	58	149
7134	F	87	2	2	2	2	999	2	Arthritic knees	1	Knee arthritis	66	149
7918	F	69	1	2	2	2	999	2	999	1	Arthrosis of the spine	73	165
7386	F	77	1	2	2	999	999	1	Hand arthritis	1	Hand arthritis	75	149
7608	F	76	2	2	2	2	999	2	999	1	Arthrosis of the knee, spine	74	157
7399	M	84	1	2	1	2	999	2	999	1	spine / knee irthrosis	94	173
	103						F	FEM.					
	Arthros is						M	BUT.					
							999	Items not asked					
	Osteop orosis.						1	Yes					
	Pac. Control						2	No					
	Smoki ng												

Printed by Books on Demand GmbH, Norderstedt / Germany